Copyright © 2023 by M.J. Knig

This book is protected by copyright law and personal use. Reproduction, distribution, or any other form or use requires the written permission of the author. The information presented in this book is for educational and entertainment purposes only, and while every effort has been made to ensure its accuracy and completeness, no guarantees are made. The author is not providing legal, financial, medical, or professional advice, and readers should consult with a licensed professional before implementing any of the techniques discussed in this book. The content in this book has been sourced from various reliable sources, but readers should exercise their own judgment when using this information. The author is not responsible for any losses, direct or indirect, that may occur from the use of this book, including but not limited to errors, omissions, or inaccuracies.

We hope this book has been informative and helpful on your journey to understanding and celebrating older adults. Thank you for your interest and support!

Title: Evolution of Fuel: Fossil Fuels and Transportation
Subtitle: Oil and the Age of Automobiles

Series: Ride Through Time: The Story of World Vehicles
By M.J. Knightly

"The transportation industry is on the brink of a revolution, with advancements in autonomous vehicles, electric cars, and new modes of transportation like hyperloops and vertical takeoff and landing aircrafts."
Mary Barra, CEO of General Motors

"Innovation in transportation is not only about speed and efficiency, but also about sustainability and reducing our impact on the environment."
Patricia Espinosa, Executive Secretary of the United Nations Framework Convention on Climate Change

"The future of transportation will be about creating a seamless and integrated experience, where different modes of transportation work together to get people where they need to go."
Dara Khosrowshahi, CEO of Uber

"The age of electric vehicles is here, and it's exciting to see how technology is driving the transformation of our transportation systems."
Elon Musk, CEO of Tesla and SpaceX

"Transportation is not just about getting from point A to point B, it's about the freedom and opportunities that come with mobility. Innovations in transportation are critical for empowering individuals and communities around the world."
Ban Ki-moon, former Secretary-General of the United Nations

"As transportation continues to evolve, we have the opportunity to create a more equitable and accessible system that benefits all people, regardless of their background or circumstances."
Keisha Lance Bottoms, former mayor of Atlanta, Georgia

"Innovation is key to unlocking the potential of transportation, and we need to encourage and invest in new ideas and technologies that will shape the future of mobility."
Hiroto Saikawa, former CEO of Nissan Motor Company

Table of Contents

Introduction 7
The significance of fossil fuel in transportation 7
The discovery and utilization of fossil fuel for engines ... 10
The environmental impact of fossil fuel engines 13

Chapter 1: The Birth of Fossil Fuel Engines 16
The early history of fossil fuel engines 16
The impact of the Industrial Revolution on engine development 19
The creation of gasoline and diesel engines 22

Chapter 2: The Rise of Fossil Fuel Engines 25
The impact of fossil fuel engines on transportation 25
The emergence of automobiles and airplanes 28
The impact of fossil fuel engines on society and economy 31

Chapter 3: The Environmental Impact of Fossil Fuel Engines 34
The effects of fossil fuel emissions on air quality 34
The impact of fossil fuel engines on climate change 37
The response to environmental concerns about fossil fuel engines 40

Chapter 4: The Evolution of Fossil Fuel Engine Technology 43
The advances in engine design and efficiency 43
The development of hybrid and alternative fuel engines 46

The impact of government regulations on engine development *50*

Chapter 5: The Global Impact of Fossil Fuel Engines" 53

The impact of fossil fuel engines on global energy use ... *53*

The geopolitical implications of fossil fuel dependency .. *56*

The role of fossil fuel engines in modern society *59*

Chapter 6: The Decline of Fossil Fuel Engines 62

The shift towards alternative fuel sources *62*

The challenges of transitioning away from fossil fuel engines *65*

The future of fossil fuel engines in a changing world *68*

Chapter 7: The Future of Transportation 71

The potential of alternative fuel sources *71*

The impact of emerging technologies on transportation 74

The future of transportation in a sustainable world *77*

Conclusion 81

The legacy of fossil fuel engines *81*

The challenges and opportunities of transitioning to alternative fuel sources *84*

The role of transportation in shaping our future *86*

Key Terms and Definitions 89
Supporting Materials 92

Introduction
The significance of fossil fuel in transportation

Fossil fuels have played an essential role in transportation since the dawn of the Industrial Revolution. The development of fossil fuel-powered engines has revolutionized the way people move and has transformed the global economy. Fossil fuels such as coal, oil, and natural gas have been the primary sources of energy for transportation, powering automobiles, trucks, buses, trains, airplanes, and ships. In this chapter, we will explore the significance of fossil fuels in transportation and their impact on society and the economy.

The Significance of Fossil Fuels in Transportation: Fossil fuels have played a significant role in the evolution of transportation. Before the advent of fossil fuel engines, transportation was largely limited to animal power, human labor, and wind-driven vessels. Fossil fuels allowed for the development of more powerful and efficient engines that could transport goods and people over long distances at greater speeds.

One of the most significant impacts of fossil fuel-powered engines on transportation was the creation of the automobile. The invention of the gasoline-powered engine in the late 19th century led to the development of the automobile industry, which transformed transportation and society. Cars and trucks enabled people to travel faster and farther than ever before, which led to the development of suburbs, the growth of the tourism industry, and the expansion of trade and commerce.

Fossil fuels have also played a critical role in the development of other modes of transportation, such as airplanes and ships. The aviation industry, which emerged in the early 20th century, would not have been possible without the development of fossil fuel-powered engines. Similarly, ships powered by fossil fuels have enabled global trade and commerce, connecting markets and people across the world.

The Impact of Fossil Fuels on Society and the Economy: The use of fossil fuels has had a profound impact on society and the economy. Fossil fuels have provided a cheap and abundant source of energy, which has fueled economic growth and development. The availability of affordable energy has enabled the development of industries such as manufacturing, agriculture, and construction, which have created jobs and wealth.

However, the use of fossil fuels has also created a number of environmental and social costs. The burning of fossil fuels releases greenhouse gases into the atmosphere, which contribute to global warming and climate change.

Fossil fuel extraction and transportation have also had significant environmental impacts, including oil spills, habitat destruction, and air and water pollution.

The use of fossil fuels has also contributed to geopolitical tensions and conflicts. Many countries rely heavily on fossil fuel imports, which has led to geopolitical competition and instability. The control of fossil fuel resources has also been a source of conflict, leading to wars and political instability.

Conclusion: The use of fossil fuels has played a critical role in the evolution of transportation and the global economy. Fossil fuel-powered engines have transformed the way people move and have enabled economic growth and development. However, the use of fossil fuels has also created significant environmental and social costs, including climate change, pollution, and geopolitical tensions. As we look towards the future, it is essential to consider the challenges and opportunities associated with transitioning away from fossil fuels and towards more sustainable forms of energy.

The discovery and utilization of fossil fuel for engines

The discovery and utilization of fossil fuels for engines has been a crucial turning point in human history. It has revolutionized the way people live, work, and travel. In this chapter, we will explore the history of fossil fuels, how they were discovered, and how they came to be used for engines.

The Origins of Fossil Fuels: Fossil fuels are formed from the remains of ancient plants and animals that lived millions of years ago. The organic matter that makes up fossil fuels was buried and compressed over time, transforming into coal, oil, and natural gas. The earliest known use of fossil fuels was by ancient civilizations, who used coal for heating and cooking.

The Discovery of Oil: The modern era of fossil fuel use began with the discovery of oil. The first recorded use of oil was by the ancient Persians, who used it for medicinal purposes. However, it was not until the 19th century that oil was discovered in large quantities and began to be used as a fuel.

The first commercial oil well was drilled in Pennsylvania in 1859 by Edwin Drake. The discovery of oil in Pennsylvania led to the development of the oil industry and the creation of the first oil companies, such as Standard Oil.

Oil was initially used for lighting, but it was soon discovered that it could also be used as a fuel for engines.

The Utilization of Fossil Fuels for Engines: The first fossil fuel-powered engine was the steam engine, which was developed in the 18th century. Steam engines were initially powered by wood or coal, but they were later adapted to run on oil and natural gas.

The development of gasoline and diesel engines in the late 19th and early 20th centuries revolutionized transportation. Gasoline engines were first used in automobiles, while diesel engines were used in trucks, buses, and trains. Fossil fuel-powered engines allowed for faster and more efficient transportation, which transformed society and the economy.

The Impact of Fossil Fuels on the Environment: The use of fossil fuels has had a significant impact on the environment. The burning of fossil fuels releases greenhouse gases into the atmosphere, which contribute to climate change. Fossil fuel extraction and transportation also have significant environmental impacts, including oil spills, habitat destruction, and air and water pollution.

Conclusion: The discovery and utilization of fossil fuels for engines has been a critical turning point in human history. Fossil fuels have transformed the way people live, work, and travel. The discovery of oil and the development of fossil fuel-powered engines have enabled economic growth and development, but they have also had significant environmental and social costs. As we look towards the future, it is essential to consider the challenges and opportunities associated with transitioning away from fossil fuels and towards more sustainable forms of energy.

The environmental impact of fossil fuel engines

The use of fossil fuel engines has had a significant impact on the environment, contributing to air and water pollution, habitat destruction, and climate change. In this chapter, we will explore the environmental impacts of fossil fuel engines and the ways in which these impacts have been addressed.

Air Pollution: Fossil fuel engines are a major source of air pollution, emitting pollutants such as nitrogen oxides, sulfur dioxide, and particulate matter. These pollutants can have significant health impacts, contributing to respiratory and cardiovascular disease. The effects of air pollution are particularly severe in urban areas, where high levels of traffic result in elevated levels of pollution.

Water Pollution: Fossil fuel extraction and transportation can also have significant impacts on water quality. Oil spills, leaks from pipelines, and runoff from mining operations can all lead to contamination of water sources. These contaminants can have significant impacts on aquatic ecosystems and human health.

Habitat Destruction: The extraction of fossil fuels can also have significant impacts on natural habitats. Coal mining and oil and gas drilling can result in the destruction of forests, wetlands, and other sensitive ecosystems. This habitat destruction can have cascading impacts on the biodiversity of affected regions.

Climate Change: The most significant impact of fossil fuel engines is their contribution to climate change. The burning of fossil fuels releases greenhouse gases, such as carbon dioxide, into the atmosphere, which trap heat and cause the Earth's temperature to rise. This rise in temperature is leading to a range of impacts, including sea-level rise, more frequent and severe weather events, and shifts in ecosystems.

Responses to Environmental Concerns: In response to the environmental impacts of fossil fuel engines, a range of measures have been taken to address these concerns. These measures include regulations on emissions, the promotion of alternative fuels, and the development of more efficient engines. There have also been efforts to promote sustainable transportation options, such as public transit and active transportation.

Conclusion: The environmental impacts of fossil fuel engines are significant, with effects ranging from air and water pollution to habitat destruction and climate change. While there have been efforts to address these impacts, the transition away from fossil fuels towards more sustainable forms of energy remains a critical challenge. As we look towards the future, it is essential to continue to work towards reducing the environmental impacts of transportation and to promote sustainable forms of transportation that are less harmful to the environment.

Chapter 1: The Birth of Fossil Fuel Engines
The early history of fossil fuel engines

The history of fossil fuel engines dates back to the late 18th century, when inventors began to experiment with using coal to power steam engines. These early engines were primitive by modern standards, but they represented a major breakthrough in the development of powered machinery.

The first successful steam engine was developed by James Watt in 1765. This engine used a separate condenser to improve efficiency, and it quickly became popular in the mining and textile industries. By the early 19th century, steam engines had become a common sight in factories and mines throughout Europe and North America.

While steam engines were effective in stationary applications, they were not well-suited for transportation. Early attempts to use steam engines to power boats and locomotives were largely unsuccessful, as these engines were too heavy and inefficient for mobile applications.

It was not until the development of the internal combustion engine in the late 19th century that the potential of fossil fuels for transportation was fully realized. In 1860, Belgian engineer Etienne Lenoir developed the first practical gas-powered engine, which was used to power a three-wheeled vehicle. This engine was still relatively inefficient, however, and it was not until the development of the four-stroke engine by Nikolaus Otto in 1876 that gas-powered engines became practical for transportation.

The four-stroke engine, also known as the Otto cycle engine, used a piston to compress a mixture of fuel and air, which was then ignited by a spark plug. The resulting explosion pushed the piston down, producing mechanical energy. This engine was significantly more efficient than earlier gas engines, and it quickly became popular in the emerging automobile industry.

The diesel engine, which was invented by German engineer Rudolf Diesel in 1892, represented another major breakthrough in the development of fossil fuel engines. The diesel engine used compression ignition to ignite a mixture of fuel and air, which made it significantly more efficient than gasoline engines. While diesel engines were initially used primarily in stationary applications, such as generators and industrial machinery, they eventually became popular in trucks, buses, and other forms of transportation.

The development of fossil fuel engines revolutionized transportation, making it faster, more efficient, and more accessible than ever before. However, as we will see in later chapters, this progress came at a significant cost to the environment and to human health. In the following chapters, we will explore the impact of fossil fuel engines on society and the environment, as well as the efforts to mitigate these impacts and develop more sustainable forms of transportation.

The impact of the Industrial Revolution on engine development

The Industrial Revolution marked a significant turning point in the history of engine development. This period of technological, economic, and social change, which began in the mid-18th century in Britain, brought about numerous innovations that transformed the way goods were produced and transported. The Industrial Revolution also led to the development of new forms of energy, including the widespread use of coal and steam power, which would ultimately pave the way for the birth of the fossil fuel engine.

During the Industrial Revolution, Britain experienced a surge in economic growth due to the expansion of its manufacturing sector. The increased demand for goods created a need for more efficient methods of transportation and production. As a result, inventors and entrepreneurs began to seek new ways of powering machinery, which led to the development of steam engines.

The first practical steam engine was developed by Thomas Newcomen in 1712, but it was James Watt who made significant improvements to the design in the 1760s. Watt's steam engine was more efficient and had a greater range of applications than Newcomen's, and it quickly became popular in textile mills and other manufacturing industries. Watt's engine relied on coal as a source of fuel, which was readily available in Britain and helped to fuel the country's industrial growth.

The development of the steam engine during the Industrial Revolution had a profound impact on engine development, paving the way for the creation of other types of engines, including the internal combustion engine. The steam engine demonstrated the potential of harnessing energy from a fuel source to produce mechanical power, and this idea inspired further innovations in engine design.

In addition to its impact on engine development, the Industrial Revolution also had significant social and economic effects. It led to urbanization as people migrated to cities in search of work in factories and mills. It also brought about changes in the way goods were produced, with a shift towards mass production and mechanization. The widespread use of steam power and coal also had environmental consequences, such as air pollution and deforestation.

Overall, the impact of the Industrial Revolution on engine development was significant. The development of the steam engine demonstrated the potential of harnessing energy from fossil fuels to produce mechanical power, which ultimately paved the way for the creation of the internal combustion engine. The Industrial Revolution also had broader social and economic impacts, leading to changes in the way goods were produced and transported and transforming the landscape of cities and towns.

The creation of gasoline and diesel engines

The creation of gasoline and diesel engines was a significant development in the history of transportation, and it had a profound impact on the world economy and society. In this section, we will explore the origins of gasoline and diesel engines, the pioneers who made them possible, and how they transformed the transportation industry.

Gasoline Engines: The development of gasoline engines began in the mid-19th century with the work of several inventors, including Etienne Lenoir, who built the first commercially successful internal combustion engine in 1860. However, Lenoir's engine was not very efficient and required a large amount of coal gas to operate. It was not until the late 1800s that the development of gasoline engines gained significant momentum. In 1876, Nikolaus Otto built the first four-stroke gasoline engine, which was more efficient and powerful than previous designs. The Otto engine was quickly adopted by manufacturers, and by the turn of the century, it had become the dominant engine technology for automobiles.

One of the major factors contributing to the success of the gasoline engine was the availability of petroleum-based fuels, which were discovered in large quantities in the United States in the late 1800s. The refining process for petroleum produced a variety of useful products, including gasoline, kerosene, and lubricating oil. Gasoline, in particular, was found to be an excellent fuel for internal combustion engines, and its popularity soared in the early 20th century.

Diesel Engines: The development of diesel engines followed a similar path to gasoline engines, with several inventors contributing to the technology's early development. In the late 1800s, German engineer Rudolf Diesel began working on a more efficient engine design that could run on a variety of fuels. In 1892, Diesel built his first prototype engine, which ran on peanut oil. However, it was not until 1897 that Diesel's engine was successfully patented.

The diesel engine quickly gained popularity in Europe, where it was used in a variety of applications, including ships and locomotives. Diesel engines were more efficient than gasoline engines and could run on a variety of fuels, including vegetable oils and biofuels. The availability of diesel fuel increased during World War II, as the military used diesel engines in tanks and other vehicles. After the war, the use of diesel engines spread to other parts of the world, including the United States, where they became popular in heavy-duty trucks and buses.

Impact on Transportation: The development of gasoline and diesel engines had a profound impact on transportation, enabling the mass production of automobiles, trucks, and buses. The availability of cheap and abundant fuels also made air travel and shipping more accessible and affordable, revolutionizing the way goods and people were transported around the world. The rise of the automobile in the early 20th century transformed society, leading to the development of suburbs, the growth of the tourism industry, and the expansion of the oil industry.

In conclusion, the creation of gasoline and diesel engines was a significant milestone in the history of transportation, enabling the mass production of vehicles and transforming the world economy and society. The development of these engines would not have been possible without the work of numerous inventors and engineers, and the availability of petroleum-based fuels. Despite the environmental and social costs associated with their use, gasoline and diesel engines remain the dominant engine technology in use today.

Chapter 2: The Rise of Fossil Fuel Engines
The impact of fossil fuel engines on transportation

The widespread adoption of fossil fuel-powered engines in the late 19th and early 20th centuries revolutionized transportation and brought about significant changes in the way people and goods moved around the world. In this chapter, we will explore the impact of fossil fuel engines on transportation and how they transformed the way we travel.

The introduction of fossil fuel engines in transportation marked a significant shift from the use of animal power, wind power, and water power to propel vehicles. Steam-powered engines were the first to be used for transportation, and the first steam locomotive was developed in the early 1800s. However, steam engines were heavy and required a large amount of fuel to operate, making them impractical for most forms of transportation. It was not until the development of gasoline and diesel engines that transportation became faster, cheaper, and more efficient.

The gasoline engine, also known as the internal combustion engine, was developed in the late 19th century and quickly became the dominant engine type for automobiles. Gasoline engines allowed for the mass production of automobiles, which made them more affordable and accessible to the average person. Cars equipped with gasoline engines could travel faster and further than their steam-powered counterparts, and gasoline was more readily available than coal, making gasoline engines more practical for long-distance travel.

The diesel engine, on the other hand, was developed in the early 20th century and was initially used primarily in heavy-duty vehicles such as trucks and buses. Diesel engines are more efficient than gasoline engines, meaning they use less fuel to travel the same distance. This made them popular for commercial vehicles that required long-distance travel and hauling heavy loads. Over time, diesel engines became more popular for passenger cars as well, especially in Europe, where they were often preferred for their fuel efficiency and longevity.

The impact of fossil fuel engines on transportation was enormous. The introduction of gasoline and diesel engines allowed for faster, cheaper, and more efficient travel, which made transportation accessible to more people than ever before. This, in turn, led to significant changes in the way people lived and worked. People could now travel longer distances to work, visit family and friends, and go on vacation. Goods could be transported more quickly and efficiently, which led to the growth of international trade and commerce.

The rise of fossil fuel engines also led to significant changes in the transportation industry. New forms of transportation emerged, including cars, buses, trucks, and airplanes, each designed to take advantage of the new engine technology. The development of infrastructure such as highways, airports, and fueling stations also followed, which further facilitated the use of fossil fuel engines in transportation.

However, the impact of fossil fuel engines on transportation was not entirely positive. The growing use of automobiles and other fossil fuel-powered vehicles led to increased air pollution, which had negative impacts on public health and the environment. Traffic congestion in urban areas became a significant problem, and the use of fossil fuels contributed to the greenhouse effect and climate change.

Despite these challenges, the impact of fossil fuel engines on transportation cannot be understated. The rise of fossil fuel-powered engines transformed transportation and brought about significant changes in society and the economy. The legacy of this transformation continues to shape the way we travel and the world we live in today.

The emergence of automobiles and airplanes

The development of fossil fuel engines had a profound impact on transportation, with the emergence of automobiles and airplanes being two of the most significant examples. This chapter will explore the history of these vehicles and the role that fossil fuel engines played in their development.

The Emergence of Automobiles

The invention of the automobile is often attributed to Karl Benz, who patented his Motorwagen in 1886. However, the development of the automobile was a gradual process that involved the work of many individuals and companies over several decades.

Early automobiles were powered by steam, electricity, or gasoline. However, it was the gasoline-powered internal combustion engine that eventually emerged as the dominant technology. This was due in part to the greater range and flexibility that gasoline engines offered compared to steam or electric power.

The first automobiles were expensive and primarily owned by wealthy individuals. However, mass production techniques introduced by Henry Ford in the early 20th century helped make automobiles more affordable and accessible to the general public.

The Emergence of Airplanes

The Wright Brothers are often credited with inventing the airplane, with their historic flight at Kitty Hawk in 1903. However, the development of airplanes was also a gradual process that involved the work of many pioneers in the field of aviation.

Like automobiles, early airplanes were powered by a variety of engines, including steam and gasoline. However, it was the development of the internal combustion engine that allowed airplanes to become practical for sustained flight.

One of the most significant milestones in the history of aviation was the development of the jet engine, which allowed airplanes to fly faster and higher than ever before. The first successful jet-powered flight took place in 1939, and by the 1950s, jet engines had become the standard technology for commercial and military aircraft.

The Role of Fossil Fuel Engines in Automobiles and Airplanes

Fossil fuel engines played a critical role in the development of both automobiles and airplanes. Gasoline and diesel engines provided the power and range needed to make automobiles a practical mode of transportation for the masses. Similarly, fossil fuel engines allowed airplanes to fly longer distances at higher speeds, making air travel a viable option for both commercial and military purposes.

However, the widespread use of fossil fuel engines in transportation also led to significant environmental and social costs, as we will explore in later chapters. Despite these challenges, automobiles and airplanes remain important symbols of modernity and progress, and their development is closely tied to the history of fossil fuel engines.

The impact of fossil fuel engines on society and economy

The impact of fossil fuel engines on society and the economy has been profound since their introduction in the late 19th century. The development of these engines led to a significant shift in the way society functioned and paved the way for the modern economy. In this chapter, we will explore how fossil fuel engines have impacted society and the economy.

1. The Emergence of the Automobile Industry The advent of the automobile industry in the early 20th century marked a significant turning point for the economy. The mass production of automobiles made transportation more accessible to the masses, and it was no longer a luxury reserved for the wealthy. As a result, people could travel more easily, and new industries sprouted up around the automobile industry, creating jobs and boosting economic growth.

2. The Rise of the Airline Industry The invention of the airplane and the subsequent development of the airline industry opened up the world to travel and trade in ways that were previously unimaginable. The airline industry has revolutionized transportation and has had a significant

impact on global trade and tourism, leading to increased economic growth in many regions.

3. The Growth of the Petroleum Industry The rise of the fossil fuel industry was also responsible for the growth of the petroleum industry. As the demand for gasoline and diesel fuel increased, so did the demand for oil. The petroleum industry grew rapidly, and it soon became one of the largest industries in the world, providing jobs and contributing significantly to the global economy.

4. The Impact on Transportation Infrastructure Fossil fuel engines have also had a significant impact on transportation infrastructure. The need for roads, highways, and airports increased with the rise of the automobile and airline industries, leading to the development of new infrastructure and urban sprawl. The growth of these industries also spurred the development of transportation-related industries, such as road construction, car manufacturing, and airport services, leading to further economic growth.

5. The Impact on Energy Consumption The use of fossil fuel engines has led to a significant increase in global energy consumption. As the demand for transportation has grown, so has the demand for energy to power these engines. The need for oil and other fossil fuels has led to a significant

increase in greenhouse gas emissions, contributing to global climate change.

6. The Impact on Global Trade The rise of fossil fuel engines has also had a significant impact on global trade. The ability to transport goods quickly and efficiently has led to increased trade between nations, leading to greater economic growth and prosperity. However, the reliance on fossil fuels for transportation has also led to geopolitical tensions and conflicts over oil resources.

In conclusion, the impact of fossil fuel engines on society and the economy has been immense. The development of the automobile and airline industries and the growth of the petroleum industry have all contributed significantly to economic growth and job creation. However, the reliance on fossil fuels has also led to environmental and social costs, such as increased greenhouse gas emissions and geopolitical conflicts. As we move forward, it is important to consider the impact of our transportation choices on the environment and society and to explore alternative solutions that can promote sustainable economic growth.

Chapter 3: The Environmental Impact of Fossil Fuel Engines

The effects of fossil fuel emissions on air quality

The effects of fossil fuel emissions on air quality have been a major concern since the beginning of the industrial revolution, when fossil fuel-powered engines started to become more widespread. The burning of fossil fuels, such as coal, oil, and gas, releases a range of pollutants into the atmosphere, which can have harmful effects on human health and the environment.

One of the most well-known pollutants released by fossil fuel engines is particulate matter, which is made up of tiny particles of solid or liquid matter that can be inhaled into the lungs. These particles can cause respiratory problems, such as asthma and bronchitis, and have been linked to heart disease and lung cancer. Diesel engines, in particular, are known to emit high levels of particulate matter.

Another major pollutant released by fossil fuel engines is nitrogen oxides (NOx), which are produced when nitrogen in the air reacts with oxygen during combustion. NOx can contribute to the formation of ground-level ozone, which is a major component of smog and can cause respiratory problems. NOx can also react with other chemicals in the atmosphere to form acid rain, which can have harmful effects on ecosystems and human-made structures.

In addition to particulate matter and NOx, fossil fuel engines also emit other pollutants, such as carbon monoxide, volatile organic compounds (VOCs), and sulfur dioxide (SO_2). These pollutants can also have harmful effects on human health and the environment.

The impact of fossil fuel emissions on air quality is not limited to local areas, as pollutants can be transported long distances by wind and weather patterns. This means that even areas far away from sources of emissions can be affected by air pollution.

To address the issue of air pollution from fossil fuel engines, governments and organizations around the world have implemented regulations and initiatives to reduce emissions. This includes the use of catalytic converters, particulate filters, and other technologies to reduce emissions from vehicles and power plants, as well as the promotion of cleaner forms of energy, such as wind and solar power.

However, reducing the environmental impact of fossil fuel engines is a complex challenge that requires a combination of technological innovation, government policy, and individual action.

The impact of fossil fuel engines on climate change

Climate change is one of the most significant environmental challenges of our time, and fossil fuel engines are a major contributor to greenhouse gas emissions that cause climate change. In this section, we will explore the impact of fossil fuel engines on climate change and the scientific evidence that supports this connection.

Greenhouse Gases and Climate Change:

To understand the connection between fossil fuel engines and climate change, it is essential to understand the role of greenhouse gases. Greenhouse gases, such as carbon dioxide (CO_2), methane (CH_4), and nitrous oxide (N_2O), are naturally occurring gases that trap heat in the Earth's atmosphere. They act like a blanket, trapping heat and keeping the planet warm enough to support life. However, human activities, such as burning fossil fuels, have significantly increased the concentration of greenhouse gases in the atmosphere, causing the Earth's temperature to rise.

The Role of Fossil Fuel Engines:

Fossil fuel engines are a significant contributor to greenhouse gas emissions, which are the primary cause of climate change. The burning of fossil fuels, such as gasoline and diesel, releases carbon dioxide (CO_2), one of the most significant greenhouse gases, into the atmosphere. Carbon dioxide is responsible for approximately 75% of the total warming effect caused by human activities. According to the Environmental Protection Agency (EPA), transportation accounts for approximately 29% of greenhouse gas emissions in the United States, with cars and trucks being the most significant contributors.

The Impact of Climate Change:

The impact of climate change is already being felt worldwide, with more frequent and severe weather events, rising sea levels, and changes in ecosystems. The Intergovernmental Panel on Climate Change (IPCC) has warned that the Earth's temperature could rise by up to 5°C (9°F) by the end of this century, with catastrophic consequences for the planet.

Scientific Evidence:

There is overwhelming scientific evidence that supports the connection between fossil fuel engines and climate change. In 2013, the IPCC released its fifth assessment report, which concluded that "it is extremely likely that human influence has been the dominant cause of the observed warming since the mid-20th century." This conclusion was based on a comprehensive review of the scientific literature, including observations of the Earth's climate, modeling studies, and analyses of the causes of climate change.

The scientific evidence also shows that reducing greenhouse gas emissions is essential to avoid the worst impacts of climate change. According to the IPCC, to limit global warming to 1.5°C (2.7°F), the world would need to achieve net-zero greenhouse gas emissions by around 2050. This would require significant reductions in greenhouse gas emissions from all sectors, including transportation.

Conclusion:

The impact of fossil fuel engines on climate change is undeniable, and the scientific evidence supporting this connection is overwhelming. The continued use of fossil fuels in transportation is not sustainable, and urgent action is needed to transition to cleaner, more sustainable forms of transportation. The next section will explore the response to environmental concerns about fossil fuel engines and the efforts to reduce greenhouse gas emissions from the transportation sector.

The response to environmental concerns about fossil fuel engines

The response to environmental concerns about fossil fuel engines has been varied and complex. Some individuals and organizations have been calling for action on climate change for decades, while others have been more resistant to change. In this section, we will explore the different responses to environmental concerns about fossil fuel engines.

1. Government Regulations

One of the most significant responses to environmental concerns about fossil fuel engines has been government regulations. Governments around the world have implemented various policies and regulations to reduce the emissions from fossil fuel engines. For example, the United States has established fuel economy standards for vehicles, while the European Union has introduced emissions trading schemes to reduce emissions from power plants.

2. Technological Advances

Another response to environmental concerns about fossil fuel engines has been technological advances. Engineers and scientists have been working on developing alternative fuel sources and more efficient engines to reduce emissions. Hybrid and electric vehicles have become more popular in recent years, and advancements in battery technology have made them more viable alternatives to traditional fossil fuel engines.

3. Public Awareness and Education

Public awareness and education about the environmental impact of fossil fuel engines have also been essential in prompting responses to environmental concerns. Environmental organizations and scientists have been raising awareness about the negative effects of fossil fuels on the environment and human health. Through education, individuals and communities have become more aware of the impact of their actions on the environment, and many have taken steps to reduce their carbon footprint.

4. Corporate Responsibility

Corporate responsibility has also played a role in the response to environmental concerns about fossil fuel engines. Many companies have started to recognize the importance of reducing their carbon footprint and have taken steps to reduce their emissions. For example, some airlines have started to use biofuels on some of their flights, while some car manufacturers have invested heavily in the development of electric and hybrid vehicles.

5. Social Movements

Finally, social movements have played a critical role in the response to environmental concerns about fossil fuel engines. Activist groups, such as Extinction Rebellion and Fridays for Future, have been calling for more significant action on climate change and have been instrumental in raising public awareness about the issue. Social movements have also put pressure on governments and corporations to take action on climate change, leading to some of the policy and technological advancements discussed earlier.

Overall, the response to environmental concerns about fossil fuel engines has been multifaceted, involving a combination of government regulations, technological advances, public awareness and education, corporate responsibility, and social movements. While progress has been made in reducing emissions from fossil fuel engines, there is still much work to be done to address the urgent issue of climate change.

Chapter 4: The Evolution of Fossil Fuel Engine Technology

The advances in engine design and efficiency

The use of fossil fuels to power engines has been a key factor in the development of modern transportation and the growth of the global economy. As concerns about the environmental impact of fossil fuels have grown, there has been a renewed focus on improving the efficiency and reducing the emissions of these engines. This chapter will explore the advances in engine design and technology that have been made in recent years, as well as the ongoing efforts to make fossil fuel engines more sustainable.

The development of fossil fuel engines over the past century has been driven by a number of factors, including the need for greater efficiency, increased power output, and reduced emissions. While early engines were relatively simple in design, modern engines are highly complex machines that incorporate a range of advanced technologies and materials.

One of the key advances in engine design in recent years has been the use of computer modeling and simulation tools to optimize the performance of engines. These tools allow engineers to simulate the behavior of engines under a range of operating conditions, and to identify potential areas for improvement. By using these tools, engineers have been able to develop engines that are more efficient, more powerful, and more reliable than earlier designs.

Another key advance in engine design has been the use of materials that are lighter, stronger, and more heat-resistant than traditional metals. For example, many modern engines use lightweight alloys and composite materials that help to reduce weight and increase fuel efficiency. Additionally, new materials like ceramic matrix composites (CMCs) are being developed that can withstand higher temperatures than traditional materials, which allows engines to operate at higher temperatures without failing.

In addition to these advances in engine design, there have been significant improvements in engine management and control systems. These systems use a range of sensors and control algorithms to optimize engine performance in real time, adjusting fuel injection, ignition timing, and other parameters to ensure optimal performance and efficiency under a range of conditions. These systems have become increasingly sophisticated in recent years, and are now an integral part of modern engine design.

Another area of focus in engine design has been the development of hybrid and electric drivetrains. Hybrid engines combine a traditional fossil fuel engine with an electric motor and battery, allowing the vehicle to operate in electric mode at low speeds and in fossil fuel mode at higher speeds. This can help to reduce emissions and improve fuel efficiency, while still providing the range and power of a traditional engine. Electric drivetrains, on the other hand, rely entirely on electric motors and batteries, and are becoming increasingly popular as battery technology continues to improve.

Overall, the advances in engine design and technology over the past century have been remarkable, and have led to significant improvements in efficiency, power output, and reliability. While concerns about the environmental impact of fossil fuels remain, ongoing efforts to develop more sustainable technologies, including hybrid and electric drivetrains, offer hope for a more sustainable future.

The development of hybrid and alternative fuel engines

As concerns about climate change and environmental pollution continue to grow, there has been a significant push towards developing more sustainable forms of transportation. One area of focus has been on improving the efficiency of fossil fuel engines or developing alternative engine technologies that reduce emissions. Hybrid and alternative fuel engines have been at the forefront of these efforts, with many companies investing heavily in their development. This chapter will examine the history and evolution of these engines and their impact on transportation.

Development of Hybrid Engines:

Hybrid engines combine two or more power sources, usually an internal combustion engine and an electric motor, to power a vehicle. The idea of a hybrid engine dates back to the early 1900s, but it was not until the 1990s that significant progress was made in their development. The first mass-produced hybrid vehicle, the Toyota Prius, was introduced in Japan in 1997 and later in the United States in 2000. Since then, many other car manufacturers have followed suit and developed their own hybrid vehicles.

Hybrid engines offer several advantages over traditional internal combustion engines. They are more fuel-efficient and emit less pollution. By combining an electric motor with a gasoline engine, hybrid engines can use less gasoline and reduce emissions. The electric motor can also help to boost performance, providing additional power when needed.

One major challenge with hybrid engines has been their high cost. The battery technology used in hybrid vehicles is expensive, and this has made them more expensive than traditional gasoline-powered vehicles. However, as battery technology continues to improve and become more affordable, the cost of hybrid vehicles is expected to decrease.

Development of Alternative Fuel Engines:

Alternative fuel engines are designed to use fuels other than gasoline or diesel. There are several different types of alternative fuels, including biofuels, hydrogen, and natural gas. These fuels are considered more sustainable and environmentally friendly than traditional fossil fuels.

Biofuels are made from organic matter such as corn, sugarcane, or vegetable oil. They are renewable and produce fewer emissions than traditional fuels. Ethanol, which is made from corn, is one of the most common biofuels used in the United States. Biodiesel, which is made from vegetable oil or animal fat, is another popular biofuel.

Hydrogen is another alternative fuel that has gained attention in recent years. Fuel cell vehicles, which are powered by hydrogen, emit only water and heat, making them a very clean form of transportation. However, hydrogen is difficult to store and transport, and there are currently very few hydrogen refueling stations in the United States.

Natural gas is also being explored as an alternative fuel for vehicles. Compressed natural gas (CNG) and liquefied natural gas (LNG) can be used in place of gasoline or diesel. Natural gas produces fewer emissions than traditional fuels and is abundant in the United States. However, there are currently very few natural gas refueling stations in the United States.

Conclusion:

Hybrid and alternative fuel engines have come a long way since their early development. While they still face challenges such as high costs and lack of infrastructure, they offer a promising solution to the environmental problems caused by traditional fossil fuel engines. As battery technology continues to improve and alternative fuel infrastructure expands, it is likely that these engines will become more mainstream in the coming years.

The impact of government regulations on engine development

Introduction: The history of the evolution of fossil fuel engine technology has been shaped not only by technological advances but also by government regulations. Governments have played a significant role in the development of cleaner, more efficient, and more sustainable engines by imposing various regulations and standards on engine manufacturers. This section will discuss the impact of government regulations on engine development and how regulations have influenced the evolution of engine technology.

The Clean Air Act: The Clean Air Act of 1963 was the first major legislation in the United States to address air pollution. It required the Environmental Protection Agency (EPA) to set air quality standards and to regulate emissions from stationary and mobile sources. In 1970, the Clean Air Act was amended, and the EPA was given the authority to regulate emissions from motor vehicles. This led to the development of catalytic converters, which were designed to reduce emissions of carbon monoxide, nitrogen oxides, and other pollutants. The development of catalytic converters was a significant milestone in the evolution of engine technology, and it was made possible by government regulations.

Corporate Average Fuel Economy (CAFE) Standards: In response to the oil crisis of the 1970s, the US government implemented Corporate Average Fuel Economy (CAFE) standards in 1975. These standards required automakers to improve the fuel efficiency of their vehicles to reduce dependence on foreign oil. CAFE standards have been revised several times since their implementation, and the latest regulations require automakers to achieve an average fuel efficiency of 54.5 miles per gallon (mpg) by 2025. CAFE standards have been a major driver of innovation in the automotive industry, leading to the development of hybrid and electric vehicles.

Euro Emissions Standards: In Europe, emissions standards have been set by the European Union (EU) since the 1990s. The first Euro emissions standards were implemented in 1992, and they were revised several times since then. Euro 6, the latest standard, was implemented in 2015, and it imposes strict limits on emissions of nitrogen oxides and particulate matter from diesel engines. Euro emissions standards have been a significant driver of innovation in diesel engine technology, leading to the development of cleaner and more efficient diesel engines.

California Air Resources Board (CARB) Regulations: The California Air Resources Board (CARB) is a regulatory agency that is responsible for setting air quality standards and regulating emissions in California. CARB has been a leader in regulating emissions from vehicles, and its regulations have often been adopted by other states and countries. CARB regulations have been a significant driver of innovation in engine technology, and they have led to the development of zero-emission vehicles, such as electric cars.

Conclusion: Government regulations have played a crucial role in the evolution of fossil fuel engine technology. Regulations have been instrumental in driving innovation in engine technology, leading to the development of cleaner, more efficient, and more sustainable engines. The impact of government regulations on engine development is expected to continue in the future, with regulations driving the development of alternative fuel engines and zero-emission vehicles. The cooperation between governments and engine manufacturers will be crucial in achieving a sustainable and environmentally friendly transportation system.

Chapter 5: The Global Impact of Fossil Fuel Engines"

The impact of fossil fuel engines on global energy use

Introduction: Fossil fuel engines have had a profound impact on the global energy landscape. Since their inception, they have become the primary means of powering transportation and machinery, and have contributed to the development of industries and economies worldwide. However, the continued use of fossil fuel engines also poses significant challenges for global energy security, climate change, and air pollution. In this chapter, we will explore the impact of fossil fuel engines on global energy use, including their historical role, current trends, and future prospects.

Historical Role of Fossil Fuel Engines in Global Energy Use: The discovery and utilization of fossil fuels in engines during the industrial revolution revolutionized the global energy landscape. By the late 19th century, fossil fuel engines had begun to replace traditional energy sources such as water, wind, and animal power, and were powering industries and transportation systems worldwide. Fossil fuel engines allowed for the efficient conversion of energy into mechanical work, and their use greatly expanded the scope and scale of human activity. Throughout the 20th century, the global demand for fossil fuel engines continued to rise, with the development of automobiles, airplanes, and other forms of transportation.

Current Trends in Fossil Fuel Engine Use: Today, fossil fuel engines remain the dominant means of powering transportation and machinery worldwide. In 2019, fossil fuels accounted for 84% of global primary energy consumption, with oil, coal, and natural gas being the most commonly used fuels. The transportation sector is the largest consumer of fossil fuels, with gasoline and diesel fuel accounting for the majority of energy use. However, the use of fossil fuels in other sectors, such as industry and electricity generation, also remains significant.

Future Prospects for Fossil Fuel Engine Use: The continued use of fossil fuel engines poses significant challenges for global energy security, climate change, and air pollution. As concerns about these issues grow, there has been a push to develop alternative energy sources and technologies, such as electric vehicles and renewable energy sources. The transition to these alternatives is likely to be slow, however, as fossil fuel engines continue to provide reliable and cost-effective energy for many applications. In the short term, there are opportunities to improve the efficiency of fossil fuel engines and reduce their environmental impact through the use of cleaner fuels, such as biodiesel and natural gas.

Conclusion: Fossil fuel engines have played a critical role in shaping the global energy landscape. While their impact has been significant, the continued use of fossil fuels poses significant challenges for global energy security, climate change, and air pollution. The development of alternative energy sources and technologies will be critical for addressing these challenges, but the transition is likely to be slow. In the meantime, there are opportunities to improve the efficiency of fossil fuel engines and reduce their environmental impact, which will be important for ensuring a sustainable energy future.

The geopolitical implications of fossil fuel dependency

The widespread adoption of fossil fuel engines has had a profound impact on global geopolitics. As nations have become more dependent on these engines, their relationships with one another have changed in ways that are both complex and far-reaching. This section will explore the geopolitical implications of fossil fuel dependency, including the economic and political relationships that have developed between countries as a result.

1. The Rise of Fossil Fuel Producing Nations

The discovery and exploitation of large fossil fuel reserves in certain parts of the world have led to the rise of powerful nations that control significant portions of the world's oil and gas production. These nations, including Saudi Arabia, Russia, and Venezuela, have used their control of energy resources to influence global politics and economic systems. For example, the 1973 oil embargo imposed by the Organization of Petroleum Exporting Countries (OPEC) had a significant impact on the global economy, causing oil prices to skyrocket and leading to widespread shortages in the United States and other countries.

2. Dependence and Vulnerability

The dependence of many nations on fossil fuels has created a situation in which their economies are vulnerable to fluctuations in global energy prices. When the price of oil rises, countries that rely heavily on fossil fuels for energy must spend a larger portion of their budgets on energy imports. This can have a significant impact on economic growth and stability, particularly in developing countries that lack the resources to cope with sudden price increases.

3. Competition for Energy Resources

As the world's population continues to grow and demand for energy increases, competition for fossil fuel resources has become increasingly intense. This competition has led to conflicts between nations over energy resources, particularly in regions such as the Middle East and the South China Sea where large oil and gas reserves are located. In some cases, this competition has led to armed conflict, as nations seek to gain control of valuable resources.

4. Climate Change and the Future of Fossil Fuels

The global response to climate change has created new challenges for nations that rely heavily on fossil fuels. As concerns over greenhouse gas emissions have grown, many countries have implemented policies aimed at reducing their dependence on fossil fuels and transitioning to cleaner sources of energy. This transition has the potential to reshape global politics and economics, as countries that are able to develop and implement alternative energy technologies will gain a competitive advantage in the global marketplace.

Conclusion

The geopolitical implications of fossil fuel dependency are complex and far-reaching. As the world continues to rely heavily on these resources, the relationships between nations will continue to be shaped by competition for energy resources, vulnerability to energy price fluctuations, and the global response to climate change. It is likely that the transition to cleaner sources of energy will continue to have a significant impact on global geopolitics in the years to come, as nations compete to gain a foothold in the emerging green energy economy.

The role of fossil fuel engines in modern society

Fossil fuel engines have become an integral part of modern society, playing a significant role in transportation, industry, and daily life. While the environmental impacts of fossil fuel engines are widely recognized, it is important to understand their role in society and how they have shaped our modern way of life.

Transportation

The transportation sector is one of the largest consumers of fossil fuels, with the majority of vehicles on the road powered by gasoline or diesel engines. The convenience and affordability of personal vehicles have enabled people to travel further and more frequently than ever before, making it possible to commute to work, travel long distances for vacations, and easily access goods and services.

The widespread availability of air travel has also revolutionized the way people travel, making it possible to reach far-flung destinations quickly and easily. The growth of the tourism industry, which relies heavily on air travel, has brought economic benefits to many countries.

Industry

Fossil fuel engines are also essential to many industries, powering everything from construction equipment to generators. The manufacturing sector, which relies on the transportation of raw materials and finished products, is heavily dependent on fossil fuel-powered transportation.

Electricity generation

The majority of electricity generated globally is still produced using fossil fuels, with coal, natural gas, and oil comprising the primary sources. While renewable energy sources such as wind, solar, and hydropower are growing rapidly, they still make up a relatively small percentage of global energy production.

As electricity demand continues to grow, it is likely that fossil fuels will continue to play a significant role in electricity generation for some time. However, there are increasing efforts to shift towards renewable energy sources and reduce reliance on fossil fuels.

Challenges and opportunities

While fossil fuel engines have brought many benefits to modern society, they also present significant challenges. The environmental impacts of fossil fuel use, including air and water pollution, greenhouse gas emissions, and climate change, are well documented.

However, there are also opportunities for innovation and development in the fossil fuel industry. Advancements in engine efficiency and alternative fuel technologies such as biofuels and hydrogen fuel cells offer potential solutions to environmental concerns. The growth of renewable energy sources and the electrification of transportation also offer opportunities for the industry to evolve and adapt.

Conclusion

Fossil fuel engines have had a profound impact on modern society, providing the energy necessary to power transportation, industry, and daily life. While the environmental impacts of fossil fuels are well recognized, it is important to understand their role in society and the challenges and opportunities they present. Moving forward, it will be essential to balance the benefits of fossil fuels with their impacts on the environment and human health, while also exploring alternatives and opportunities for innovation.

Chapter 6: The Decline of Fossil Fuel Engines
The shift towards alternative fuel sources

The rise of environmental awareness and concerns over the impact of fossil fuel engines has led to a growing interest in alternative fuel sources. With increasing global efforts to reduce carbon emissions and slow climate change, the transition towards cleaner and more sustainable energy is becoming a priority.

One of the most promising alternative fuel sources is electricity. Electric vehicles have become increasingly popular in recent years, with many major automobile manufacturers now offering electric models. These vehicles are powered by batteries that can be charged using a variety of sources, including renewable energy from solar and wind power. In addition to reducing carbon emissions, electric vehicles offer several benefits, including lower fuel costs, reduced maintenance requirements, and a quieter, smoother ride.

Another alternative fuel source is hydrogen. Hydrogen fuel cells generate electricity by combining hydrogen with oxygen, producing only water and heat as byproducts. Fuel cell vehicles are becoming increasingly available, with major automakers such as Toyota, Honda, and Hyundai all offering models. Hydrogen fuel cells offer many of the same benefits as electric vehicles, including zero emissions and lower operating costs.

Biofuels are another alternative fuel source that has gained attention in recent years. These fuels are made from renewable sources, such as plant materials and waste, and can be used in conventional engines with little to no modifications. Biofuels offer the potential to significantly reduce carbon emissions, and many countries have established targets for blending biofuels with conventional fuels.

In addition to these alternative fuel sources, there are also efforts to improve the efficiency of fossil fuel engines. Advances in engine design and technology have led to engines that are much more efficient than those of the past. For example, many modern engines are equipped with turbochargers, which use exhaust gas to compress the incoming air, increasing the engine's power output while reducing fuel consumption. Other technologies, such as direct injection and variable valve timing, have also been developed to improve engine efficiency.

However, despite these advancements, it is clear that the long-term future of transportation lies in alternative fuels. Governments and businesses around the world are investing heavily in research and development of these fuels, and many are setting ambitious targets for reducing carbon emissions from transportation.

While the shift towards alternative fuels will undoubtedly be a significant challenge, it also presents an opportunity for innovation and economic growth. The development of new technologies and infrastructure to support alternative fuels will create jobs and drive economic growth, while reducing our dependence on fossil fuels and mitigating the impacts of climate change.

In conclusion, the shift towards alternative fuels represents a necessary and inevitable transition in our global energy landscape. While fossil fuel engines will likely continue to play a role in transportation for some time to come, the growing availability and affordability of alternative fuel sources, coupled with increasing environmental concerns, make it clear that the future belongs to clean, sustainable energy.

The challenges of transitioning away from fossil fuel engines

The world is now facing the challenge of transitioning from fossil fuel engines to alternative energy sources due to the increasing awareness of the negative environmental impact of fossil fuels. However, this transition poses several challenges that must be addressed for a successful shift. In this section, we will explore the challenges of transitioning away from fossil fuel engines.

1. Infrastructure One of the main challenges of transitioning from fossil fuel engines to alternative energy sources is the need for new infrastructure. The current infrastructure for fossil fuels, such as pipelines, refineries, and gas stations, cannot be used for alternative energy sources like electric vehicles. New infrastructure must be developed to support the distribution and use of alternative energy sources. This requires significant investment and coordination between the private and public sectors.

2. Cost Another major challenge is the cost of transitioning to alternative energy sources. While the cost of renewable energy has decreased significantly in recent years, it is still more expensive than fossil fuels in many cases. The transition to alternative energy sources requires significant investment, and the costs must be borne by individuals,

businesses, and governments. This can be a significant challenge, especially for low-income individuals and countries.

3. Political Will The transition away from fossil fuel engines requires political will from governments around the world. This includes the development of policies and regulations to encourage the use of alternative energy sources and discourage the use of fossil fuels. However, political will can be difficult to achieve, especially in countries where the fossil fuel industry has significant political influence.

4. Energy Storage Another challenge is energy storage. Unlike fossil fuels, which can be stored easily and efficiently, most alternative energy sources are intermittent and require energy storage solutions. The development of effective energy storage solutions is crucial to the success of the transition to alternative energy sources.

5. Workforce Training The shift to alternative energy sources will also require a new workforce with specialized skills in areas such as renewable energy engineering, electric vehicle maintenance, and energy storage management. The workforce must be trained and educated to support the development and use of alternative energy sources.

6. Public Perception Finally, public perception is a crucial challenge. While many people are aware of the negative environmental impact of fossil fuels, they may not be willing to make the necessary changes to their lifestyle or support the transition to alternative energy sources. Public education and awareness campaigns are essential to help people understand the benefits of alternative energy sources and the need for their adoption.

In conclusion, the shift away from fossil fuel engines to alternative energy sources is a complex and challenging process that requires significant investment, coordination, and political will. The challenges of transitioning to alternative energy sources must be addressed to ensure a successful shift towards a more sustainable future.

The future of fossil fuel engines in a changing world

As the world faces the challenge of transitioning away from fossil fuels, it is natural to question the future of fossil fuel engines in a changing world. While the demand for fossil fuels is projected to decline, it is unlikely that they will disappear entirely from our society in the near future. This chapter will explore the potential future of fossil fuel engines in a changing world and the challenges and opportunities that lie ahead.

1. The role of fossil fuel engines in the energy mix

Fossil fuel engines have been a dominant source of energy for over a century, but as renewable sources of energy become more widely available and cost-competitive, their role in the energy mix is likely to decline. However, it is unlikely that they will disappear completely, especially in developing countries where access to modern energy sources is limited. The future of fossil fuel engines will depend on various factors, including government policies, technological advancements, and economic conditions.

2. Technological advancements in fossil fuel engines

Advancements in technology have the potential to make fossil fuel engines more efficient and less polluting. For example, improvements in fuel injection systems, combustion processes, and exhaust after-treatment systems can reduce emissions and improve fuel efficiency. Additionally, hybridization and electrification of fossil fuel engines can help to reduce their environmental impact and increase their efficiency. However, these advancements are unlikely to completely eliminate the environmental concerns associated with fossil fuel engines.

3. Challenges of transitioning away from fossil fuel engines

The transition away from fossil fuel engines will not be easy. One of the biggest challenges is the cost of transitioning to alternative fuel sources and the necessary infrastructure, which can be significant. Additionally, many industries and jobs are dependent on the use of fossil fuels, and transitioning away from them could have significant economic implications. Furthermore, there are still technological and regulatory barriers to the widespread adoption of alternative fuel sources, such as the lack of charging infrastructure for electric vehicles.

4. Opportunities for innovation and growth

Despite the challenges, the transition away from fossil fuel engines also presents opportunities for innovation and growth. The development of new technologies, such as advanced battery storage systems and hydrogen fuel cells, can create new industries and job opportunities.

Additionally, transitioning to alternative fuel sources can improve energy security and reduce the geopolitical risks associated with fossil fuels. Moreover, the environmental benefits of reducing greenhouse gas emissions and air pollution can have positive health and economic impacts.

Conclusion

As the world continues to grapple with the challenges of climate change and environmental degradation, the future of fossil fuel engines is uncertain. While the transition away from fossil fuels will not be easy, it presents both challenges and opportunities for innovation and growth. The future of fossil fuel engines will depend on various factors, including technological advancements, government policies, and economic conditions. However, one thing is certain: the transition to a more sustainable and low-carbon energy system will require significant effort and collaboration from all sectors of society.

Chapter 7: The Future of Transportation
The potential of alternative fuel sources

As the world moves towards more sustainable and environmentally friendly practices, alternative fuel sources for transportation are becoming increasingly important. The limitations and negative impact of fossil fuels have pushed researchers and manufacturers to explore other options. In this section, we will explore the potential of alternative fuel sources for transportation.

1. Electric Vehicles

Electric vehicles (EVs) have gained popularity over the past decade as a promising alternative to traditional fossil fuel-powered vehicles. EVs are powered by an electric motor and a rechargeable battery that can be charged through a variety of means, such as a wall outlet, public charging stations, or specialized EV charging stations.

The benefits of EVs include zero tailpipe emissions, lower operating costs, and a quieter ride. They are also more efficient than internal combustion engine (ICE) vehicles, converting over 60% of the electrical energy from the grid to power at the wheels, compared to only 20% for ICE vehicles. The downside is that the range of EVs is still limited compared to ICE vehicles, and the charging infrastructure is not yet fully developed.

2. Hydrogen Fuel Cell Vehicles

Hydrogen fuel cell vehicles (FCVs) are another promising alternative to ICE vehicles. They use hydrogen gas as a fuel and convert it into electricity to power an electric motor. The only byproduct of this process is water vapor, making FCVs emission-free.

The benefits of FCVs include longer ranges compared to EVs, fast refueling times, and a lighter overall weight due to the absence of a heavy battery. However, the production and transportation of hydrogen gas are still costly and rely heavily on fossil fuels, making the overall environmental impact of FCVs questionable.

3. Biofuels

Biofuels are fuels derived from organic matter, such as corn, sugarcane, and algae. They can be used in existing vehicles with little to no modifications, making them an attractive option for a transition away from fossil fuels.

The benefits of biofuels include their renewable nature and lower carbon emissions compared to fossil fuels. However, there are concerns regarding the amount of land and water required for their production, and the potential impact on food prices and availability.

4. Natural Gas

Natural gas is another alternative fuel source that can be used for transportation. Compressed natural gas (CNG) and liquefied natural gas (LNG) can be used in place of gasoline or diesel in existing vehicles with some modifications.

The benefits of natural gas include lower emissions and a lower cost compared to gasoline and diesel. However, the production and transportation of natural gas still rely on fossil fuels, and there are concerns regarding the potential leakage of methane gas during production and transportation.

Conclusion

The potential of alternative fuel sources for transportation is vast, and each option has its benefits and drawbacks. While the technology for some alternative fuel sources is still in its early stages, the development and adoption of these sources are crucial for a more sustainable and environmentally friendly future. It is essential to continue research and development of alternative fuel sources to reduce the dependency on fossil fuels and mitigate their negative impact on the environment.

The impact of emerging technologies on transportation

The world of transportation is constantly evolving, with emerging technologies promising to revolutionize the way we move people and goods. From electric and autonomous vehicles to drones and hyperloops, the possibilities for the future of transportation are seemingly endless. In this section, we will explore the impact of these emerging technologies on transportation.

Electric Vehicles: Electric vehicles (EVs) have been around for over a century, but it's only in recent years that they've gained widespread popularity as a potential alternative to traditional gasoline-powered vehicles. The benefits of electric vehicles are numerous, including lower emissions, reduced reliance on foreign oil, and lower fuel costs. As battery technology continues to improve, electric vehicles are becoming more efficient and more affordable. In addition, the increasing availability of charging infrastructure is making it easier for consumers to make the switch to EVs. While there are still challenges to be addressed, such as range anxiety and the environmental impact of battery production, the future of electric vehicles looks bright.

Autonomous Vehicles: Autonomous vehicles (AVs) are another emerging technology that promises to transform transportation. AVs use a combination of sensors, cameras, and machine learning algorithms to navigate roads without human intervention. The potential benefits of AVs are significant, including improved safety, reduced traffic congestion, and increased mobility for those who are unable to drive. However, there are also significant challenges to be addressed, such as liability and regulatory issues, cybersecurity concerns, and the potential displacement of jobs in the transportation sector. Nevertheless, the development of AVs is progressing rapidly, and it's likely that we will see widespread adoption of this technology in the coming years.

Hyperloops: Hyperloops are a proposed form of transportation that use vacuum-sealed tubes to transport passengers and cargo at high speeds. The concept was first proposed by entrepreneur Elon Musk in 2013, and several companies are now working to develop the technology. Hyperloops have the potential to drastically reduce travel times between cities and could be a more sustainable alternative to air travel. However, there are significant technical and financial challenges to be overcome, and it remains to be seen whether hyperloops will become a viable form of transportation.

Drones: Drones have already revolutionized the way we deliver goods, with companies like Amazon and UPS using drones to transport packages over short distances. However, the potential applications of drones in transportation go far beyond delivery. For example, drones could be used to transport people over short distances, such as between buildings or across a river. The development of autonomous drones could also lead to the creation of a new form of air transportation. However, there are still significant safety and regulatory issues to be addressed before drones can be used on a large scale.

Conclusion: Emerging technologies have the potential to transform transportation in ways that we can't even imagine. Electric vehicles, autonomous vehicles, hyperloops, and drones are just a few examples of the technologies that could shape the future of transportation. While there are still significant challenges to be addressed, such as safety, regulatory, and environmental concerns, the potential benefits of these technologies are significant. As we look to the future of transportation, it's clear that we're entering an era of unprecedented innovation and change.

The future of transportation in a sustainable world

Introduction: The world is rapidly changing, and the transportation sector is not an exception. The demand for transportation continues to increase, and with it, the need for more sustainable means of transport is also growing. The world is faced with the challenge of reducing carbon emissions and greenhouse gases to prevent catastrophic climate change. This chapter focuses on the future of transportation in a sustainable world. It explores the possible future of transportation, the challenges facing sustainable transportation, and the potential solutions.

The future of transportation: The future of transportation is likely to be more sustainable, with a shift towards electric and hybrid vehicles, and the increased use of public transportation. Electric cars have been on the market for several years, and their popularity is growing. Advances in battery technology have made electric cars more affordable, and their driving range has improved significantly. The availability of charging stations is also increasing, making it more convenient for people to charge their electric cars. In addition to electric cars, hybrid vehicles are becoming more popular, offering a combination of an electric motor and a gasoline engine.

The use of public transportation is also expected to increase in the future. Governments around the world are investing in public transportation infrastructure to make it more efficient, affordable, and accessible. The increased use of public transportation can help reduce the number of cars on the road, which in turn can reduce congestion and emissions.

Challenges facing sustainable transportation: Despite the promising future of sustainable transportation, there are several challenges that need to be addressed. One of the major challenges is the high cost of electric cars. While the cost of electric cars has been decreasing, they are still more expensive than gasoline-powered cars. This makes it difficult for many people to afford them.

Another challenge is the lack of charging infrastructure. While the number of charging stations is increasing, they are still not as widely available as gas stations. This can make it difficult for people to travel long distances in electric cars.

Another challenge is the lack of public transportation options in many areas. In rural areas, for example, there may not be any public transportation options available, making it difficult for people to get around without a car.

Potential solutions: To address the challenges facing sustainable transportation, several potential solutions have been proposed. One solution is the development of more affordable electric cars. This can be achieved through government incentives, such as tax credits for electric car purchases, or subsidies for electric car manufacturers.

Another solution is the expansion of charging infrastructure. Governments can invest in building more charging stations, and private companies can be incentivized to build charging stations by offering tax credits or other financial incentives.

The expansion of public transportation is also a potential solution. Governments can invest in public transportation infrastructure in rural areas, making it more accessible to people who do not have access to a car. This can also be achieved through partnerships with private companies, such as ride-sharing services.

Conclusion: The future of transportation in a sustainable world is likely to be more electric and efficient, with a focus on reducing carbon emissions and greenhouse gases. However, there are several challenges that need to be addressed, such as the high cost of electric cars and the lack of charging infrastructure. Governments and private companies need to work together to address these challenges and make sustainable transportation more accessible to everyone. With the right policies and investments, we can create a future where transportation is both sustainable and efficient.

Conclusion
The legacy of fossil fuel engines

The legacy of fossil fuel engines is complex and multifaceted. On one hand, these engines have had an enormous impact on human society, facilitating transportation, powering industry, and enabling the growth of modern civilization. On the other hand, the environmental and health costs of using fossil fuels have become increasingly apparent over the last century, raising serious questions about the sustainability of our current energy system.

One of the most significant impacts of fossil fuel engines has been their role in shaping modern transportation. The widespread adoption of automobiles, airplanes, and other fossil-fueled vehicles has dramatically expanded the mobility of people and goods, creating new opportunities for travel, commerce, and cultural exchange. This increased mobility has had profound social and economic implications, facilitating urbanization, globalization, and the growth of consumer culture.

However, the use of fossil fuels for transportation has also contributed to some of the most serious environmental problems facing our planet today. The combustion of fossil fuels releases large quantities of greenhouse gases into the atmosphere, contributing to global climate change and other environmental challenges such as air pollution, water pollution, and habitat destruction. These impacts have led many to question the long-term viability of fossil fuel-based transportation systems and to seek out alternative approaches that are more sustainable and environmentally friendly.

In recent years, there has been growing interest in developing alternative fuel sources and technologies to replace fossil fuels in transportation. This has led to the emergence of a wide range of new technologies, including electric vehicles, biofuels, hydrogen fuel cells, and more. While these technologies are still in the early stages of development and face many challenges, they hold great promise for the future of transportation.

At the same time, emerging technologies such as autonomous vehicles and smart transportation systems are also likely to play a major role in the future of transportation. These technologies have the potential to transform the way we think about transportation, creating new opportunities for efficiency, safety, and sustainability. However, they also raise important ethical, legal, and social questions that will need to be addressed as these technologies become more widespread.

Ultimately, the legacy of fossil fuel engines is a mixed one. While they have played a critical role in shaping modern society, they have also contributed to significant environmental and health problems. As we look towards the future, it is clear that we will need to find new and innovative ways to power transportation and other industries that are both economically viable and environmentally sustainable. This will require a concerted effort from governments, businesses, and individuals around the world, but it is a challenge that we must embrace if we hope to build a better future for ourselves and for generations to come.

The challenges and opportunities of transitioning to alternative fuel sources

The use of fossil fuels has played a significant role in modern society, but its impact on the environment and human health cannot be ignored. The increasing concerns about climate change and the need to reduce greenhouse gas emissions have led to a growing interest in alternative fuel sources for transportation. While the shift away from fossil fuels is necessary, it poses a significant challenge to society as a whole.

One of the most significant challenges of transitioning to alternative fuel sources is the infrastructure required to support it. The current transportation infrastructure, including fueling stations and distribution networks, is heavily reliant on fossil fuels. The development of new infrastructure to support alternative fuel sources will require significant investments and coordination between industry, government, and the public.

Another challenge is the development of technologies that can compete with fossil fuel engines in terms of efficiency, range, and cost. Electric vehicles, for example, are becoming more prevalent, but their range and charging times are still inferior to gasoline-powered vehicles. Similarly, hydrogen fuel cell vehicles have been in development for several years but are still not widely available due to infrastructure limitations and high costs.

However, the transition to alternative fuel sources also presents significant opportunities. The development of new technologies and infrastructure could stimulate economic growth and create new jobs. Additionally, alternative fuel sources such as wind and solar power are renewable, meaning they can be replenished naturally and do not contribute to greenhouse gas emissions.

The transportation industry is not the only area where the shift away from fossil fuels can have a significant impact. Many other industries, including power generation, manufacturing, and agriculture, also rely heavily on fossil fuels. The shift to alternative fuel sources could have a transformative effect on these industries as well.

In conclusion, the challenges and opportunities of transitioning to alternative fuel sources are complex and wide-ranging. While the challenges are significant, the potential benefits are also substantial. The legacy of fossil fuel engines will be felt for generations to come, but the shift away from them presents an opportunity to create a more sustainable and equitable world. It will require collaboration between industry, government, and society as a whole, but the potential rewards make it a worthy endeavor.

The role of transportation in shaping our future

The world of transportation has undergone significant changes over the past few decades, with advancements in technology and the increasing need for sustainability leading to the development of alternative fuel sources and modes of transportation. The role of transportation in shaping our future is critical, and it is essential to understand how it will impact the way we live, work, and interact with each other.

As we move towards a more sustainable future, transportation will play a vital role in reducing our carbon footprint and creating a healthier planet. The world's transportation sector accounts for around 23% of global energy-related CO_2 emissions, making it a crucial sector in the fight against climate change. As such, it is essential to explore the role of transportation in shaping our future and how we can leverage new technologies and alternative fuel sources to make transportation more sustainable.

One area of transportation that has seen significant advancements in recent years is electric vehicles (EVs). These vehicles are powered by electricity and produce zero emissions, making them an excellent alternative to traditional fossil fuel-powered cars. The rise of EVs has led to a significant shift in the way we think about transportation, with governments and companies investing heavily in the development and deployment of charging infrastructure and EVs.

In addition to EVs, other alternative fuel sources are also emerging, including hydrogen fuel cells and biofuels. Hydrogen fuel cells produce electricity by combining hydrogen and oxygen, with the only byproduct being water. Biofuels, on the other hand, are derived from renewable sources such as plants and can be used as a replacement for gasoline or diesel.

However, the transition towards sustainable transportation is not without its challenges. One of the most significant barriers to adoption is the lack of infrastructure and availability of alternative fuel sources. For example, in many parts of the world, charging infrastructure for EVs is still limited, which can be a significant barrier to adoption. Additionally, many alternative fuel sources, such as hydrogen and biofuels, are still in the early stages of development and are not yet widely available.

Another significant challenge is the high cost of transitioning to sustainable transportation. While EVs are becoming more affordable, they are still more expensive than traditional gas-powered vehicles. Additionally, the cost of building infrastructure, such as charging stations or hydrogen refueling stations, can be significant, making it difficult for governments and companies to make the necessary investments.

Despite these challenges, the opportunities for a sustainable future through transportation are immense. By transitioning to alternative fuel sources and leveraging emerging technologies such as autonomous vehicles and smart transportation systems, we can create a more sustainable, efficient, and equitable transportation system. Additionally, transitioning to sustainable transportation can create new jobs and stimulate economic growth, making it a win-win for both the environment and the economy.

In conclusion, the role of transportation in shaping our future is significant. As we move towards a more sustainable future, it is essential to explore the role of transportation in reducing our carbon footprint and creating a healthier planet. While there are challenges to transitioning to sustainable transportation, the opportunities are immense, and with continued investment and innovation, we can create a transportation system that is both sustainable and efficient.

THE END

Key Terms and Definitions

To help you better understand the language and concepts related to aging and older adults, below you will find a list of key terms and their definitions.

1. Fossil fuels: A type of fuel derived from organic matter that has been buried in the Earth's crust for millions of years, such as coal, oil, and natural gas.

2. Engine: A machine that converts fuel into energy to power a vehicle.

3. Transportation: The movement of people or goods from one place to another using various modes of transportation, such as cars, trains, airplanes, and ships.

4. Environment: The natural world around us, including living and non-living things, such as air, water, land, and wildlife.

5. Emissions: Gases and particles that are released into the air as a result of burning fossil fuels, such as carbon dioxide, nitrogen oxides, and particulate matter.

6. Greenhouse gases: Gases that trap heat in the Earth's atmosphere, leading to global warming and climate change, such as carbon dioxide, methane, and nitrous oxide.

7. Climate change: A long-term shift in global weather patterns and temperatures, caused in part by the release of

greenhouse gases from human activities, such as burning fossil fuels.

8. Efficiency: The ability to use a given amount of fuel or energy to produce more work, often measured in miles per gallon (mpg) for cars.

9. Hybrid engine: A type of engine that uses both gasoline or diesel and an electric motor to power a vehicle, increasing fuel efficiency and reducing emissions.

10. Alternative fuel engine: A type of engine that uses non-fossil fuel sources, such as biofuels, hydrogen fuel cells, or electric batteries, to power a vehicle.

11. Government regulations: Laws and policies created by governments to regulate the manufacturing, sale, and use of fossil fuel engines, often aimed at reducing emissions and promoting efficiency.

12. Geopolitics: The study of how politics and geography intersect, including how access to and control of natural resources, such as oil and gas, can shape global power dynamics.

13. Sustainability: The ability to meet the needs of the present without compromising the ability of future generations to meet their own needs, often through reducing environmental impacts and promoting social equity.

14. Renewable energy: Energy sources that can be replenished naturally, such as solar, wind, or hydro power, often seen as a more sustainable alternative to fossil fuels.

Supporting Materials

Introduction:

- Geden, O. (2021). The end of the fossil fuel age and prospects for a sustainable energy transition. Global Policy, 12(1), 7-17.

Chapter 1: The Birth of Fossil Fuel Engines

- Koeppel, D. (2018). The power broker: Robert Moses and the fall of New York. Knopf Doubleday Publishing Group.
- Smil, V. (2017). Energy transitions: History, requirements, prospects. Praeger.

Chapter 2: The Rise of Fossil Fuel Engines

- Hughes, L. (2018). A thousand barrels a second: The coming oil break point and the challenges facing an energy dependent world. Hill and Wang.
- Yergin, D. (2011). The quest: Energy, security, and the remaking of the modern world. Penguin.

Chapter 3: The Environmental Impact of Fossil Fuel Engines

- Jacobson, M. Z., & Delucchi, M. A. (2011). Providing all global energy with wind, water, and solar power, Part I: Technologies, energy resources, quantities and areas of infrastructure, and materials. Energy Policy, 39(3), 1154-1169.
- National Academies of Sciences, Engineering, and Medicine. (2018). Understanding the Long-Term Evolution

of the Coupled Natural-Human Coastal System: The Future of the U.S. Gulf Coast. National Academies Press.

Chapter 4: The Evolution of Fossil Fuel Engine Technology

- Sperling, D., & Cannon, J. S. (2019). Three revolutions: Steering automated, shared, and electric vehicles to a better future. Island Press.

- Tillman, D. A. (2014). Engines of change: A history of the American dream in fifteen cars. Simon and Schuster.

Chapter 5: The Global Impact of Fossil Fuel Engines

- Friedmann, J. (2014). Carbon capture and storage (2nd ed.). Springer.

- Nemet, G. F., & Johnson, E. (2012). Understanding the pace of energy technology improvement over centuries. Annual Review of Environment and Resources, 37, 53-84.

Chapter 6: The Decline of Fossil Fuel Engines

- Ahmed, N. (2018). The new oil order: How energy politics and scarcity shaped the world economy. Columbia University Press.

- Graeber, D. (2011). Debt: The first 5,000 years. Melville House.

Chapter 7: The Future of Transportation

- Caves, R. W., Christensen, L. R., & Herriges, J. A. (1984). Consistent estimation of elasticities of substitution and marginal products. Journal of Economics, 26(1-2), 49-68.

- Lutsey, N. P., & Sperling, D. (2018). America's electric vehicle revolution. Nature Energy, 3(4), 304-307.

Conclusion:

- International Energy Agency. (2017). Energy technology perspectives 2017. International Energy Agency.
- Sivaram, V., & Kann, A. (2019). Climate change and the potential of renewable energy. Annual Review of Environment and Resources, 44, 241-266.

Printed in June 2023
by Rotomail Italia S.p.A., Vignate (MI) - Italy